MEN-AT-ARMS S

EDITOR: MARTIN W.

The Sudan Campaigns 1881-1898

Text by ROBERT WILKINSON-LATHAM

Colour plates by MICHAEL ROFFE

OSPREY PUBLISHING LIMITED

Published in 1976 by
Osprey Publishing Ltd,
59 Grosvenor Street, London, W1X 9DA
© Copyright 1976 Osprey Publishing Ltd

Reprinted 1979, 1980, 1981, 1982, 1983, 1984 (twice),
1985, 1986, 1987, 1988, 1989, 1990

ISBN 0 85045 254 6

Filmset by BAS Printers Limited, Wallop, Hampshire
Printed in Hong Kong.

The British Army of the 1880's

When the British army intervened in Egyptian affairs in 1881 to smash the revolt of the Turko-Egyptian army under Arabi Pasha, it was still in the throes of the greatest reorganisation in its history. The regimental system, the time-honoured practice of denoting a regiment's seniority by its number in the Line, had been swept away with the stroke of a pen. No longer would a regiment be known by its number (and perhaps its supplementary title, many of which were given in 1782) but solely by its new 'Territorial Title'. Regimental individuality was also lost as the new system – commonly called the Cardwell reforms but really following the recommendations of the Stanley Committee – amalgamated two numbered line regiments to form the first and second battalions of each 'Territorial' regiment. In many cases this marriage was resented by both regiments, since a number of the mergers effected seemed to be without reason. As George Bernard Shaw wrote: 'The British soldier can stand up to anything except the British War Office.' The only regiments not affected were those numbered 1 to 25 inclusive, which already had two battalions; and also the 79th Highlanders. The main idea behind the reforms was that in each regiment one battalion would serve overseas on a tour of duty while the other remained for home defence, trained recruits and provided drafts for the battalion abroad. In addition, each regiment was also now allocated a fixed depot and a recruiting area. In fact, the use of county titles for a number of regiments was propitious since it gave the soldiers a permanent home and the civilians a sense of pride in their county regiment.

Besides the formation of these new regiments with titles in place of numbers, the cherished facing colours on the collars and cuffs of tunics were standardised. No more the grass green facings of the 24th, or the gosling green of the 5th Fusiliers; gone for ever the purple of the 56th and the yellow of the 57th. In their place, it was dictated that all Royal Regiments would wear blue facings, all English and Welsh regiments would wear white, Scottish regiments yellow, and Irish regiments green. Now the only distinction between, say, English non-Royal regiments lay in the regimental name embroidered on the shoulder strap, since even regimental-pattern buttons had been done away with in the early 1870s.

The British army was like most other European armies of the late nineteenth century except that it was a volunteer and not a conscripted force. As Alfred de Vigny wrote, 'An army is a nation within a nation . . .' and of the British army this

The interior of Fort Mex, Alexandria after the British naval bombardment, July 1882. British soldiers are inspecting the damage done to the defences and artillery, while warships can be seen standing-off in the distance. (Isabel and Aline Scott-Elliot)

3

was undoubtedly true. Wolseley said in his *Soldier's Pocket Book* that 'The soldier is a peculiar being that can alone be brought to the highest efficiency by inducing him to believe that he belongs to a regiment that is infinitely superior to the others round him.' While the armed forces stood apart from normal society, being considered as de Vigny wrote ' . . . one of the vices of our age', they were as Kipling was quick to point out ' . . . no thin red 'eroes, nor we aren't no blackguards too. But single men in barricks, most remarkable like you. . .' The life was hard and disciplined and the pay abysmally low, but the regiment gave to its members a home and – in peace time at least – food and lodging, in return for which the 'Soldiers of the Queen' were expected to maintain the 'Pax Britannica'.

The basic pay of the private was a shilling a day, 'the Queen's shilling', out of which stoppages had formerly been made for uniform and food.

However, a series of reforms led to the introduction of free food by 1870; this resulted in the soldier receiving every penny of his shilling, although beer money was discontinued.

In 1876, each man was given 2d. a day more, but this was termed deferred pay and was only payable on discharge as a lump sum to help the ex-soldier settle into civilian life. 'Good Conduct badges', as they were called, carried an extra 1d. a day after three years' exemplary service, but this period was later reduced to two years. Service, as in other volunteer armies, had originally been for twenty-one years, but in 1870 a system of twelve years' service was introduced by Secretary of State for War, Edward Cardwell. Even in 1882 there were still a certain number of 'old soldiers' left in the army; William Robertson, who later rose from the ranks to become a Field-Marshal, wrote in 1877 that 'The system introduced by Mr. Cardwell under which men enlisted for twelve

Officers of the Queen's Own Cameron Highlanders, Egypt 1882. Note the various patterns of 'frock' tunics, some with breast pockets and some without, also the different ways of wearing the 'Sam Browne' belt and equipment. (Isabel and Aline Scott-Elliot)

years' regular service, had not yet had time to get into full swing. Regiments were, therefore, still composed mainly of old soldiers who, although very admirable comrades in some respects and with a commendable code of honour of their own, were in many cases addicted to rough behaviour, heavy drinking, and hard swearing. They could not be blamed for this. Year in and year out they went through the same routine, were treated like machines – of an inferior kind – and having little prospect of finding decent employment on the expiration of their twenty-one years' engagement, they lived only for the present, the single bright spot in their existence being the receipt of a few shillings – perhaps not more than one – on the weekly pay-day.'

Barrack life was harsh and often insanitary. Men lived, slept and ate in their barrack room, the Government allowing 1 lb. of bread and 12 oz. of meat per day per head. Groceries, vegetables and other extras were provided by the men themselves. Bennet Burleigh, war correspondent of the *Daily Telegraph*, described what rations were issued to the soldier on campaign in the Sudan in 1898: 'He usually has a "grand appetite" when campaigning. On active service the Government ration allowed him comprises (inclusively) – bread $1\frac{1}{4}$ lb., meat $1\frac{1}{4}$ lb., tea $\frac{1}{2}$ ounce, sugar $2\frac{1}{2}$ ounces, salt $\frac{1}{2}$ ounce, rice $\frac{1}{2}$ ounce, pepper 1/36th ounce, fresh vegetables $\frac{1}{2}$ lb., or in lieu of latter, 3 ounces onions, daily. That was what he got at Dekesh. For fresh vegetables he received onions 3 ounces of. In addition he could purchase, by payment out of his own pocket, one-third more of each article. As a matter of fact, he constantly bought food, a grateful country not even giving its troops in the field sufficiently varied diet. Nowadays the commonest folk at home look for something more than plain bread and tough meat. The meat ration issued was poor, and ran largely to

Officers and men of the Egyptian army parading in marching order with packs, Cairo 1882. The summer white uniform is shown. In winter a blue uniform was worn. (Isabel and Aline Scott-Elliot)

'Friendlies'. Sudanese tribesmen friendly to Egypt, with Egyptian army officers mounted on camels, 1882. (Isabel and Aline Scott-Elliot)

bone – ¼ lb. of meat to 1 lb. of bone. In the regimental canteens, cheese, tinned milk, jams, sardines, bacon, tinned fruits, tea, coffee, date pudding, soup, etc. were sold in large quantities to the men. Were a smart contractor to take up the job, the War Office and the country might, on those lines, succeed in making campaigns pay for themselves. I commend the suggestion to them. At any rate, they would recover every farthing of the soldier's pay, and a trifle over.' Wellington had described the army private as 'the scum of the earth' in the early 1800s, and by 1880 the service had not attracted many recruits of a better calibre. True, there was selection on joining, but most of the men enlisted only because they lacked a trade or were starving. No wonder the large industrial slums of Manchester, Birmingham, Glasgow and London yielded the majority of the rank and file.

The reforms of the 1870s had done much to better the lot of the soldier, but there was still a long way to go. Officers such as Sir Garnet Wolseley realised the amateur aspect of the British army compared with the professional spirit shown by the Prussians. Up until 1871 officers had purchased commissions and promotion; Cardwell put a stop to this practice, but even so officers were still not educated in their profession. During the 1882 campaign Sir Garnet complained that 'I have seen splendid battalions kept in the rear while others of inferior quality were sent to the front because the general commanding did not dare employ against the enemy a corps whose commanding officer was manifestly incompetent . . . I hold that it is criminal to hand over in action the lives of gallant soldiers to men who are deplorably ignorant of the elements of their profession . . .'

As for the rank and file, the Rev. G. J. Hardy in his book *The British Soldier* (1915) described the types of recruits that were daily accepted. 'When trade is bad we get good recruits and when good, bad ones. The army is still recruited mainly from the class of manual labour . . . Only 49 recruits in a thousand can be described as well educated.'

If the officers were unprofessional and the rank and file recruited from the illiterate and starving, the backbone of the army was, as throughout history, 'the Non-commissioned man'. Junior officers and soldiers alike depended on the harsh judgment, skill and devotion to duty of the non-commissioned officers, the army professionals who – often of intimidating countenance – were a breed unto themselves. Everything in their lives was done 'by the book', yet they were the mainstay of each and every regiment. A regiment with good NCOs was an efficient piece of military machinery. The soldier was not encouraged to think; this was done by the non-commissioned officer who was the vital link between the rank and file and the officers. During the 1880s and 1890s the private soldier, the shilling-a-day man, was solely required to act as a mindless brick in a human wall, and the system of drill ensured that the 'wall' would stand against anything. Discipline had been maintained by the lash, loss of pay and confinement. In 1868 flogging was declining in the army, and in 1881 an Act of Parliament abolished it.

At the end of a campaign, the soldier probably received a medal with 'bars' to denote his participation in various battles and actions, but once a war was over it was back to 'peace-time soldiering' and boring garrison life.

In Egypt and the Sudan in the 1880s the British soldier was dressed as he always had been in red or scarlet, the only concession to heat being the white-covered cork helmet. Sweltering in his 'grey-back' shirt and scarlet frock, encumbered with straps of white buff leather equipment, and carrying a rifle (the Martini-Henry) that kicked like a mule, British troops began an association with Egypt and the Sudan that was to endure until the Anglo-French expedition of 1956 was aborted by pressure from the United States.

The Anglo-Egyptian Army

The Anglo-Egyptian army that fought Mahdism in the Sudan, at first on its own but later with the aid of British troops, was formed after the Arabi revolt of 1881–2. The previous army was a Turco-Egyptian force in which the British had no say. The new Egyptian conscripted army was at first limited to 6000 men with twenty-five British officers, but was later increased to over 18,000 men and 140 white officers. It consisted of eighteen battalions of infantry; six of these, numbered 9 to 14, were made up of Sudanese blacks, whose terms of service differed from those of the Egyptians. Each infantry battalion was divided into six companies of 100–120 men, giving a total battalion strength – with band and stretcher-bearers – of between 650 and 750 men.

Of the Egyptian or *fellah* battalions, 1 to 4 and 15 to 18 were officered by the British, while battalions 5 to 8 were led by so-called 'native officers'. British officers usually numbered three for a *fellah* battalion and four for a Sudanese battalion. The native officers were usually Turkish,

The charge of the Royal Horse Guards at Kassassin, September 1882, by Seccombe. This painting shows the equipment carried by cavalry during the campaign. (Parker Gallery)

Circassian or Albanian, but there were also a few Egyptians. In the Sudanese battalions there were usually a few native captains or subalterns, but as G. W. Steevens tells us, ' . . . lack of education keeps them from higher grades.'

No British officer held a lower rank than that of Major or *Bimbashi*, and matters were so arranged that there was never a native officer senior to a British one in the same battalion. Command of a battalion usually fell to a Lieutenant-Colonel or *Kaimakam*, but he was usually addressed by the courtesy title of Bey. Battalion commanders were usually captains or majors in the British army, and *Bimbashis* were subalterns.

Each battalion also had a 'Sergeant Whatsisname', as Kipling affectionately called him – a British non-commissioned officer whose task was to drill and make soldiers out of the raw material at his disposal. The NCOs, either colour sergeants or sergeants, were volunteers like the officers.

The uniform of the infantry battalions and other arms of the service was a brown jersey, sand-coloured trousers and dark blue puttees. Head wear was the tarbush with a cover, the Egyptian battalions having a neck flap in addition. They were armed with the Martini-Henry rifle and long socket bayonet.

All the cavalry were Egyptian conscripts and most of their squadron leaders were British officers. As G. W. Steevens wrote in *With Kitchener to Khartoum*, the reason for 'all-Egyptian' cavalry was that ' . . . a black can never be made to understand that a horse needs to be groomed and fed.' The cavalry consisted of ten squadrons, each numbering about 100 men.

The Egyptian artillery had two batteries of field artillery armed with Maxim-Nordenfeldt quick-firing 9-pounders or 18-pounders with a double shell, ' . . . handy little creatures which a couple of mules draw easily.' The horse battery was armed with 12-pounder Krupp guns, and the other two field batteries with 9-pounders. Again, all the gunners were Egyptian conscripts and the battery commanders British.

Finally, there was the 800-strong camel corps, divided into eight companies composed of half Sudanese and half Egyptian troops with five white officers. There were also the usual non-combatant services.

The conscripted Egyptian soldier or *fellah*, representing one in every 500 of the population, was required to serve for six years with the colours and a further nine with the reserve or police. His pay was a piastre a day (equivalent to $2\frac{1}{2}$d. in 1898) which Steevens gleefully wrote was ' . . . a magnificent salary, equal to what he would usually be making in full work in his native village.' The black Sudanese soldier was liable to be enlisted where found, and served for life. He was paid a basic 14 shillings a month to begin with and a family allowance of $3\frac{3}{4}$d. per day for those who had permission to marry. Unaccustomed to garrison and town life, the Sudanese battalions were usually quartered on the frontier. Many of the Sudanese recruits were former enemies, the better prisoners and deserters being enlisted into one of the black battalions.

Valentine Baker Pasha, an ex-officer of the 10th Hussars who had been court-martialled and dismissed from the service over his alleged conduct with a woman in a railway carriage, was offered the post of Commander-in-Chief of the new Anglo-Egyptian army, but at the last moment (some say through the intervention of Queen Victoria) the offer was withdrawn. Baker was then given the command of a ramshackle police force, the Egyptian 'Gendarmerie'.

The Dervish Army

The original Dervish army that did very much as it pleased in the Sudan during 1881–4 under the command of the Mahdi, the spiritual and temporal leader long awaited by Mohammedans, was very different from the army formed by his successor, the Khalifa, which was finally destroyed at Omdurman in 1898.

Mohammed Ahmed Ibn Al-Sayid Abdullah was the son of a boat-building carpenter, and was born in 1844. His father claimed that he was descended from the Prophet, and in 1861 Mohammed Ahmed became a Summaniya Dervish, a member of a strict Moslem sect. The word *Dervish* means a Moslem friar vowed to poverty

'The British Square'. Officers and men of the Queen's Own Cameron Highlanders in the traditional infantry square formation. Note the red puggarees around the men's sun helmets, also worn by the officers with the addition of a regimental pattern badge on the front. (Isabel and Aline Scott-Elliot)

and austerity, and the self-proclaimed Mahdi demanded these virtues in the followers – or 'ansars', as he preferred to term them – whom he rallied to his cause.

Their original 'uniform', if it can be called that, was the *jibbah*, a plain cotton garment; later, as E. N. Bennet, war correspondent of the *Westminster Gazette* pointed out in his book, *The Downfall of the Dervishes* '. . . the Mahdi, who was somewhat ascetic – in theory, at any rate, if not in practice – ordered his followers to sew black patches upon their nice white coats, as tokens of humility. But alas for human frailty, what was intended to curb the spiritual pride of the faithful became a direct incentive to the vainglorious adornment of their persons! The ladies of Omdurman were strongly opposed to the dowdiness of the black patches upon their husbands and lovers, and, under the influence of the more aesthetic circles of Dervish society, the white *gibbehs* were gradually tricked out with gaudy squares of blue, red and purple.'

The initial religious frenzy which enabled the Mahdi to defy the Egyptian army, and to massacre and defeat it, abated when he died in June 1885, some months after the death of General Gordon – 'Gordon of Khartoum', who was killed defending that city. The successor to the Mahdi was the Khalifa, Abdullah the Taiaishi, a chief of the Baggara tribe who had been considered the Mahdi's right hand as early as 1883.

The Khalifa reformed the followers into an organised army upon European lines. War correspondent Bennet Burleigh described this reformation: 'He to a great extent replaced the mad fury of the early dervishes by the introduction of military organisation among the wild tribes, endeavouring, though in a crude way, to adopt the system of training and tactics employed in the Egyptian army. Of late he has succeeded in so far modifying the original tribal system of conducting warfare, that his infantry, cavalry, and artillery are ordered and commanded much in the European fashion. The emirs and lesser leaders nowadays wear distinctive insignia showing their rank; and, more wonderful still, the restless Baggara cavalry have been dragooned, and made to drill and work by squadrons. It is, however, in the handling of military supplies and keeping the accounts of stores that the Mahdists have apparently not only sensibly copied, but bettered, the instruction of the Khedival commissariat department. Some of the dervish accounts, probably kept by Coptic clerks, which fell into my hands at Hafir, Dongola, and elsewhere, showed that, down to the uttermost pound of beans or packet of small-arms ammunition, nothing was issued without a written warrant, and that receipts for everything were taken and the stores

9

on hand could be ascertained at a glance.'

The Dervish army was divided into four separate parts, each to perform a specific duty. Firstly there were the mounted horsemen who were all Baggara tribe Arabs; then camelmen, composed of Danagla and Jaalin tribesmen. The rest of the army consisted of foot soldiers: the Jehadia, who were Sudanese blacks (with a few Arabs) armed with rifles, and the swordsmen and spearmen who were commanded by Emirs and organised into *Rubs* equivalent to battalions.

Discipline was hard in the Dervish army; smoking, drinking, wearing fine clothes and jewellery, festivities, dancing and bad language were all met with severe punishments which included flogging up to 1000 lashes. Rations were issued on a strict basis and given out in *ardebs* equivalent to 5·6 bushels. Arabs received $\frac{3}{24}$ per month, Jehadia $\frac{4}{24}$ and Baggara $\frac{6}{24}$.

Each portion of the army was subdivided into sections with section leaders; they were required to parade at regular intervals, but there was no battle training. The inborn skill and resources of the Dervish made him a first-class fighting soldier, one of the best with which 'Tommy Atkins' had brushed. As Rudyard Kipling, the unofficial spokesman for the British soldier, wrote in his famous poem '*Fuzzy Wuzzy*':

'We've fought with many men across the seas,
An' some of 'em was brave an' some was not;
The Paythan an' the Zulu an' Burmese;
But the Fuzzy was the finest o' the lot.'

Charles Neufeld, a German trader who had ventured into the Sudan in 1885 in search of gum-arabic and been captured by the Khalifa, wrote in the account of his twelve years of captivity, *A Prisoner of the Khaleefa*, about the efficiency and unbeatable fighting qualities of the Dervish. 'At close quarters the dervish horde was more than a match for the best drilled army in Europe. Swift and silent in their movement, covering the ground at four or five times the speed of trained troops, every man, when the moment of attack came accustomed to fight independently of orders, lithe, supple, nimble as cats and as bloodthirsty as starving man-eating tigers, utterly regardless of their own lives, and capable of continuing stabbing and jabbing with spear and sword while carrying half a dozen wounds, any of which would have put a European *hors de combat* – such were the 75,000 to 80,000 warriors which the Khalifa had already ... Artillery, rifles and bayonets would have been of little avail against a horde like this rushing a camp by night.'

Egypt and the Sudan

'Surely enough "When Allah made the Sudan", say the Arabs, "he laughed". You can almost hear the fiendish echo of it crackling over the fiery sands,' wrote G. W. Steevens, famous war correspondent of the *Daily Mail*. The events which occurred in the Sudan were very closely allied to Egyptian affairs, and to see how Britain became involved in Egypt one must look briefly back to the beginning of the nineteenth century.

For hundreds of years Egypt had been under the domination of foreign rulers; Arabs, Mamlukes and finally the Turks who engulfed it in the vast Ottoman Empire. By the beginning of the nineteenth century this once vast empire had started to decay rapidly and various powers, especially Russia, were watching it with more than passing interest.

Britain first became interested in Egypt in 1798 when the French, under Napoleon, invaded the country. Ever conscious of protecting India, the British government was frightened that the young Frenchman might make a bid to attack their Eastern empire by the overland route. In British eyes, the overland route to India depended almost entirely on Egypt and the eastern Mediterranean, previously considered to be of little strategic importance. The route to India had always been via the Cape of Good Hope, but Britain now decided to adopt preventive measures to guard the alternative entry. Malta and Gibraltar were regarded as being of great strategic significance and became first-line naval bases. The Turkish rule of Egypt also suffered under this new-found importance.

During the early years of the nineteenth century Viceroys of the Sultan of Turkey were appointed, the first being Mohammed Ali, a colourful rogue dubbed 'The father of modern Egypt'. He had in his former years been a harsh ruler, but his skill and wit had enabled him to gain a certain amount of autonomy from the Sultan of Turkey in Egyptian affairs, even though he was unable to read or write until he was forty.

The importance of Egypt was apparent to Britain when shortly after the accession of the Viceroy Mohammed Ali a certain British officer, Lieutenant Waghorn, organised an 'overland' route connecting the Mediterranean and the Red Sea. This venture flourished and soon a regular route was establishing. Ships came from Britain to Alexandria where they discharged their passengers and cargo; these were then transported by Waghorn's efficient river boats and baggage animals to Suez, where passengers and goods alike were put aboard Bombay-bound vessels. This 'short cut' took at least four weeks from the journey to India. To Britain, Egypt had assumed a vital importance.

Meantime, Mohammed Ali had not been content with ruling in Egypt. In 1821 he cast an eye southwards to the substantial lands lying below Egypt – the Sudan. Having conquered the northern part of this hostile and barren country by force, he found an untapped source of first-class fighting men, slaves and ivory. For the Egyptian economy this new-found wealth proved of immeasurable value. During the fifteen years between 1860 and 1875, more than 400,000 Sudanese were captured and sold by the Arab slave traders. As supplies of potential slaves in one part of the country dried up, so the Arabs moved to another area, and as a result some of the many Sudanese tribes ceased to exist.

Besides the slavers, Egyptian troops had 'colonised' the area, or at least set up garrisons to help enforce Egyptian rule. Taxes were collected by harsh and brutal means, and corruption on the part of the Egyptian officials was rife. The rhinoceros-hide whips wielded by the tax-collectors not only extracted the last few piastres, they also started something of more serious consequence. It was the British who were later to reap the harvest of years of Egyptian misrule.

By the time of the Crimean War in 1854, Said Pasha had inherited the Viceregal throne and

The battle of Tel-el-Kebir, September 1882. 'First in the Fray' by F. Dadd. Men of the Cameron Highlanders storming the earthworks constructed by Arabi Pasha and bayoneting the gunners. (National Army Museum/*Illustrated London News*)

opened the country to European traders and experts. Amongst these was the Frenchman Ferdinand de Lesseps who, inspired by Waghorn's idea, elaborated it to consider the possibility that a canal could be cut connecting the Mediterranean and Red Sea, shortening still further the route to the East. Said Pasha, always open to the persuasive tongue of the Frenchman, who was neither engineer nor builder, agreed to allot him large areas of land, free labour and mineral rights plus permission to realise his lifetime's dream. Britain, for obvious reasons, opposed the canal project in order to protect her interests in India, and the Foreign Office went so far as to warn the Government that 'At present India is unattackable. It will no longer be so when Bombay is only 4,600 miles from Marseilles.' Because of the Crimean War, in which Britain and France were allied, little pressure could be brought to bear against the project, however, and the idea went ahead. At the crucial point, as digging commenced, Said Pasha died; financial crises arose and under the forcibly re-negotiated terms of the new Viceroy, Ismail Pasha, work ceased.

After the battle of Tel-el-Kebir. The arrival of Lord Wolseley and his staff at the bridge of Tel-el-Kebir while prisoners are collected and soldiers slake their thirst. Painting by Elizabeth Butler. (Parker Gallery)

The company formed by de Lesseps claimed compensation and the arbitrator, no less a person than Napoleon III, awarded £3 million damages. To meet this bill Ismail Pasha agreed to forfeit his profits from Egypt's shares in the canal to that amount, in return for immunity from further claims. Egypt's national debt had increased from £3¼ million in 1841 to a staggering £94 million by 1876, mainly to Britain, France, Russia and other European countries.

On 17 November 1869, at a £2-million ceremony, the Suez Canal was opened, but during the next few years Egypt's economy declined rapidly. The assumption of the title of Khedive by Ismail Pasha alone cost the country £1 million which was paid to the Sultan of Turkey. By 1875 the foreign bankers who had financed Egypt began to worry and in that same year Britain unwittingly cemented herself to Egypt and the Sudan.

The major user of the Canal had been Britain, who, having been unable to prevent its construction, endeavoured to bring it under her control. In an unprecedented move Prime Minister Disraeli borrowed £4 million from the Rothschild banking family to purchase the shares of the Khedive, which represented 44 per cent of the capital. On 25 November 1875, Disraeli wrote to Lady Bradford informing her of his action, '. . . a State secret' he wrote, 'certainly the most important of the year.' Secrecy had been uttermost in the Prime Minister's plans because as he wrote, 'The day before yesterday, Lesseps, whose Company has the remaining shares, backed by the French Government, whose agent he was, made a great offer. Had it succeeded, the whole of the Suez Canal would have belonged to France, and they might have shut it up! We have given the Khedive 4 millions sterling for his interest, and run the chance of Parliament supporting us. We could not call them together for the matter, for that would have blown everything to the skies, or to Hades.' The news was received with overwhelming applause by Britain and other countries. Queen Victoria wrote that 'what she liked most was, it was a blow at Bismarck', while the King of the Belgians hailed it as 'the greatest event of modern politics, Europe breathes again . . .'

Through this move Britain had increased her interest in Egyptian affairs. In April the following year her involvement deepened when the bankrupt Egyptian government suspended payments of interest and debts. Britain and France, as the two imperial powers to whom owed most, imposed on Ismail a Commission of the Debt, a type of 'receivership' which in effect gave them control of

Egyptian affairs. In 1879, after an abortive revolt stirred up by Ismail, the British and French demanded that the Sultan of Turkey should depose him. In his place Tewfik, Ismail's son, was appointed Khedive.

The Rise of Mahdism and the Arab Revolt

While Britain and France exercised 'Dual Control' of Egypt in the affairs of government, they certainly did not contemplate expensive military intervention. Egypt's main problem had always been the Sudan and the Khedive's government had appointed various outsiders to try to manage it. The title of Governor-General of the Sudan was held by a succession of foreigners, one of whom was General Charles Gordon. By 1880 Egypt had nearly 40,000 troops in garrisons throughout the Sudan, imposing her corrupt rule through the Governor-General in Khartoum. Poverty, oppression and disaffection were rife in the province, which was constantly ravished by slave and ivory traders and plundered by the soldiers and corrupt tax collectors. Gordon, who had restored some form of just rule during his term of office, plainly saw the trouble that was brewing for Egypt. In 1880 he wrote, 'If the liberation of slaves is to take place in 1884 (in Egypt proper) and the present system of government goes on there cannot fail to be a revolt of the whole country. But our government will go on sleeping till it comes and then have to act *à l'improviste*'.

In 1881 the obscure son of a carpenter proclaimed himself the 'Mahdi' or 'Guided One of the Prophet', the long-expected Messiah of the Islamic faith. Mohammed Ahmed Ibn Al-Sayid Abdullah could not have chosen a better time. Egypt was in financial chaos, and the grip of the Turco-Egyptian army on the Sudan was weakened by foreign intervention and general apathy. The Governor-General of the Sudan was then Rauof Pasha, an incompetent and corrupt official whom

Gordon had twice dismissed from subordinate positions. The deputation sent by Rauof Pasha to the Mahdi had little effect, and the Governor-General decided to send a punitive expedition to capture the rebel leader. An ill-armed force sailed up the Nile in the steamer *Ismailia*, reaching it goal at Aba after dark. Instead of waiting until dawn the troops disembarked in chaos in the dark, and stumbling amongst the mud and reeds of the shoreline, fell easy prey to the Mahdi's *ansars*. A few Egyptians escaped to the steamer, which hastily fled.

Three months later, in December 1881, a force of 1400 Egyptian troops under Rashid Bey, Governor of Fashoda, was ambushed and hacked to pieces by the Mahdi's forces. In Egypt it

The British entry into Cairo, September 1882. Cameron Highlanders passing in review before Lord Wolseley. (Isabel and Aline Scott-Elliot)

appeared that the Mahdi in the Sudan was getting out of hand and that the Army seemed incapable of putting down the rebellion. The Mahdi's forces were swelling daily. Recruits who were eager, after the defeats of the Egyptians, to cast off corrupt and oppressive rule joined in their thousands ready to carry on the Holy War he had proclaimed.

Egypt had, however, her own internal problems brought about by the 'Dual Control' of Britain and France. As in other spheres of Egyptian rule, the key and senior posts were held by foreigners and not Egyptians. The army was run by Turkish-Circassian officers, and Egyptian officers had little say. Ahmed Bey Arabi, son of an Egyptian village chief, felt the time had come to air the grievances of many officers like himself. Army pay had been cut back, foreign domination was on the increase and the corrupt rule of the Turks in

The Military Police in Cairo dressed in scarlet tunics and white helmets with their distinctive arm-bands. The Police were armed with short carbines and the 1879 pattern artillery sword-bayonet. They are seen checking papers, keeping order and deterring beggars. (Parker Gallery)

power was devastating an already ruined country. Arrested by the Government, Arabi was brought before a Council of Ministers for censure, but troops and officers loyal to his cause burst in, turned out the Council (having emptied inkpots over them) and forced Tewfik to accept Arabi as a minister in the new government.

Feeling in Britain and France was acute. The prospect of revolution and the probability of the national debt being refuted stirred Britain into action. There was no joint plan by the two powers, since intense rivalry prevented any agreement or understanding being reached over Egypt. By May 1882 the Admiralty had ordered the Mediterranean Squadron to Alexandria and the French had also despatched some warships. Turkey, however, remained inactive despite de-

mands from both Britain and France to intervene. Arabi refused to be intimidated and started to reinforce his seaward defences, to man the forts with heavy artillery and to dig emplacements. A British ultimatum to dismantle the forts was ignored, and the Royal Navy was left to intervene alone when the French fleet sailed away because of a change in government. A bombardment commenced at 7 a.m. on 11 July 1882, and shelling between ship and shore lasted the entire day, until both parties fell silent through lack of ammunition. After two days of inactivity, watching the Egyptians firing Alexandria, looting the town and killing Christians, a party of 'Bluejackets' and Marines was ordered ashore. Their swift progress through Alexandria and the restoration of law and order were vividly described by war correspondents of the leading newspapers. Looters were summarily dealt with, either being shot on sight, or, if caught, tried by a military court and hanged as a warning to others.

The rebellion was far from over, however. With 60,000 men, Arabi was still in control of Cairo and a large part of Egypt. 'Our only General', Sir Garnet Wolsely, was despatched to Egypt to deal with the rebellious Egyptian army. In typical Wolseley fashion, he made it known to the press and other officers that he was going to attack Aboukir, but this was purely a diversionary move. Wolseley had himself written in his famous *Soldiers Pocket Book* (1869) concerning the news-hungry war correspondents that '. . . this very ardour for information a General can turn to account by spreading fake news among the gentlemen of the press and thus use them as a medium by which to deceive the enemy.' In this case, Wolseley did just that. Even de Lesseps, worried about damage to his Canal, telegraphed Arabi that the British were landing at Aboukir. This was the last message sent, as Wolseley's troops landed at Suez and closed the telegraph office. Surprised by the unexpected direction of the attack, Arabi reinforced his lines at Tel-el-Kebir.

On 28 August the Egyptian army attacked a force of about 2000 men under General Graham at Mahsama, and despite the fears of Lieutenant-General Willis (who telegraphed to Wolseley '. . . Fear Graham has been defeated') the Egyptian force, which outnumbered the British five to one, was driven off; much of the credit for the victory went to Drury Lowe's cavalry and their famous so-called 'moonlight charge'. On 10 September Wolseley's force was marching towards Arabi's well-positioned lines at Tel-el-Kabir, and the General was worried. In confidence, he wrote to his wife about his anxieties and his weak position. '. . . I have determined to move out from here on Tuesday night to attack the enemy's fortified position on Wednesday morning a little before daybreak. I am so weak that I cannot afford to indulge in any other plan, and it requires the steadiest and the best troops to attain my object – and then I may fail – oh God grant I may not! – I know that I am doing a dangerous thing, but I cannot wait for reinforcements; to do so would kill the spirit of my troops, which at present is all I could wish it to be. I hope I may never return home a defeated man: I would sooner leave my old bones here than go home to be jeered at . . . Everything depends upon the steadiness of my infantry. If they are steady in the dark – a very crucial trial – I must succeed. Otherwise I might fail altogether, or achieve very little.'

The Egyptians had done a good job of manning their lines. In addition to the 25,000 troops there were about seventy field guns including some of the latest Krupp breechloaders. The area in front of the fortifications was desert – poor fighting ground for troops attacking against earthworks. For four days Wolseley's staff reconnoitred, mapped and discussed the position, and at last discovered the 'Achilles heel' of the Egyptians: they did not man their outposts at night. Wolseley was decided: it would have to be a night attack, or failing that a night march followed by a swift dawn attack. He chose the latter course. Night marches were always risky; in the desert sound carried for miles, troops lost their way in the dark, directions were hard to follow and the sense of direction erratic. Besides these factors, the most important was that soldiers tended to become unsteady when they lost contact with their officers or comrades in the dark, and this could provoke panic or chaos and jeopardise the attack.

Progress over the desert would be slow, at about one mile per hour, therefore Wolseley timed his troops' departure for 1.30 a.m. calculating to reach the enemy's line just before dawn. Directing poles had been placed in the sand by the Royal Engineers to show the line of march, but these proved of little use. At various points during the advance incidents occurred which might have jeopardised the entire operation. Riders coming with instructions from the staff were mistaken for Arabs, although no shots were fired; and the line was disrupted when a Highland regiment rested for twenty minutes. Since this regiment was in the centre of the line, and orders were passed by word of mouth, the flanks continued to advance until they also halted, the entire line now forming a crescent with the opposing flanks confronting each other. In the dark, each could have easily mistaken the other for the enemy and opened fire, but fortunately calm prevailed.

Frank Power, *The Times* war correspondent, summed up the attack in his despatch. 'There was no moon, and thus almost within cannon shot, the

The Tokar Expedition. Disembarkation of Parker Pasha's troops with stores at Trinkitat for the relief of Tokar. The expedition was short-lived and ended in the disastrous first battle of El Teb (4 February 1884). Drawing by Melton Prior, war artist for the *Illustrated London News*. (National Army Museum/*Illustrated London News*)

two armies were resting peacefully, the one side dreaming probably little of the terrible scene of the awakening, when their rest at length rudely disturbed, they awoke to see swiftly advancing upon every side an endless line of dreaded red-coats, broken by the even *more fearful* blue of the Marines.' The Egyptians were totally defeated at a cost of 399 casualties among the British force, 243 of which were from the front-line troops of the Highland Brigade.

After a forced march the following day the troops entered Cairo and captured Arabi, who surrendered his sword to General Drury Lowe. Arabi was tried in December 1882 and banished to Ceylon, but was pardoned in 1901.

The Sudan 1881-1883

While the Egyptian army had been preoccupied with fighting the British expeditionary force, the Mahdi in the Sudan did very much as he pleased. Britain had been unwillingly drawn into Egypt, which she now garrisoned, but Gladstone's firm opposition to imperialism for whatever reasons drew the line at intervention in the Sudan. The undisputed fact was, however, that Egypt's security rested on keeping the Sudan subdued, and Britain had assumed the responsibility of Egypt. The newly-formed Anglo-Egyptian army under British officers, financed by the Khedive and not the British tax payer, would have to solve the problem. Soldiers under British officers should be able to deal with the ill-armed savages of the Sudan, who, it was thought, possessed mainly spears and a few small-arms.

The Mahdi's troops had already defeated an Egyptian force under Yussif Pasha, who succeeded Raouf Pasha in March 1882. The Mahdi had also decided to establish his base at El Obeid and by September his troops were ready for the assault. The attack launched on 8 September failed, and the Mahdi was hastily forced to find an excuse for his followers, to whom he had previously declared

The battle of Tamai, by G. D. Giles. During this battle the Dervishes 'broke' the British square. Note the fierce hand-to-hand fighting on the left, the medical orderlies attending to wounded in the centre and the reserve ammunition mules to the right. (National Army Museum)

that the enemy bullets could not kill them. Lacking adequate firepower, he settled down to starve out the city, which fell on 19 January 1883, yielding up large supplies of arms and munitions. An Egyptian relief force of 3,000 which set out in September was systematically slaughtered.

In February, Cairo learned of the disaster of El Obeid and the ensuing slaughter, and decided that some firmer action must be taken. A retired Indian army officer re-employed as Chief of Staff in the new Egyptian army was chosen to lead an expedition. William Hicks, or 'Hicks Pasha' as he was called, had not had a particularly disting-uished career. He had spent most of his career in India, having fought in the Mutiny and taken part in the Abyssinian campaign of 1867. Aided by a few other British officers but hampered by the interference of an eighty-year-old Pasha to whom he was subordinate, he endeavoured to instil some fighting spirit into his 9000 men. It was a

hopeless task. Most of them were recruited from the army defeated by Wolseley at Tel-el-Kebir, and their morale was as non-existent as their fighting qualities. This noticeable inferiority of his troops was not echoed by some of his fellow officers. 'We were all in high spirits,' wrote Colonel the Hon. J. Colborne, 'and eagerly look-ing forward to the campaign.' British aid was out of the question as the force set off towards Khartoum. 'Whether Hicks falls or conquers,' stated the *Pall Mall Gazette*, 'is not our business, not a single British soldier will be ordered to Khartoum if the Mahdi were to rout the whole force under the orders of the Khedive's officers.'

Later Colborne noted ominously in his book, *With Hicks Pasha in the Sudan* (1884), regarding the quality of the troops: 'During their passage from Cairo, men and officers had completely forgotten their drill. When the guns were attempted to be brought into action, dire confusion reigned. Men ran against each other; the ground was strewn with cartridges; hoppers were placed anywhere but where they should have been. No one ap-peared to have the slightest knowledge of how to feed, aim, and discharge the pieces. In the midst of all this, poor Walker – not knowing anything of

Sudan 1884: British troops in tropical dress in camp. Note the shapes of the tents, and the sentries guarding stores on the left. See also the Highland pattern of cutaway tunic, although the man on the left wears the standard infantry pattern. (Isabel and Aline Scott-Elliot)

the language beyond the words of command – stood aghast. General Hicks thundered out that he had never seen such a disgraceful scene in his life, and ordered Forestier-Walker to remain for three days perpetually drilling his men in that sandy scorching camp, instead of returning with us to the comparatively "blest abode" of Khartoum.'

The column moved out from Khartoum and on 26 June scored a minor victory by defeating an attack of Baggara cavalry. Three days later a more determined attack was beaten off. Colborne described with enthusiasm the start of the battle. 'Onward they came, waving their banners . . . but the Khedive's troops, encouraged by their English officers, had no fear. They had seen the charm-protected enemy bite the dust under their fire . . . But Nordenfeldts and Remingtons are no respectors of creeds.'

On reaching Jebel-Ain, Hicks thought the campaign over as there was no sign of the enemy. The next garrison he visited was that of Dueim

which had been attacked on 23 August but had beaten off the Dervishes, inflicting over 4,000 casualties. Spending almost a month there, Hicks moved south-west on 23 September on his route towards El Obeid. Morale was low, since expected reinforcements had not appeared and the camels and horses were dying at an alarming rate. 'The ill-fated army scarcely met a living soul, but flocks of vultures followed them as if waiting for their prey.' From deserters, the Mahdi knew the disposition of Hicks's force, its low morale, its lack of water and its depleted numbers owing to death and desertion. He despatched a letter to Hicks inviting him to surrender, but this was naturally ignored. Previous warnings left by the Mahdi in the form of leaflets were used by the Egyptian troops as lavatory paper.

On 3 November 1883 the remainder of Hicks's force, now down to 7,000, reached Kashgeil, twelve miles south of El Obeid. Fighting started on the 3rd and lasted until the 5th when the Mahdi's troops finally killed the last of the Khedive's force. The Dervishes had attacked Hicks's square on the 3rd and the night of the 4th. Desperately short of water, three smaller squares were formed on the morning of the 5th to get to the next waterhole.

None reached their goal. According to statements made by the Dervishes, Hicks was one of the last to die: '... he had emptied his revolver and, holding his sword in his right hand, waited for the rush of the enemy; he was soon surrounded and his horse wounded in the back; he then dismounted and fought most gallantly with his sword until he fell, pierced by several spears ...'

While the Mahdi was finishing off Hicks's column Osman Digna, an ex-slave trader whom the Mahdi had created an Emir and granted the title of a provincial governor, had defeated another Egyptian force. Osman Digna had concentrated his efforts on the eastern side of the Sudan but mainly around the area east of the Nile and the towns of Tokar and Suakin.

Having been repulsed by Tewfik Bey at Suakin, Osman Digna turned his attention to Tokar, which he besieged. A relief force of 500 accompanied by Commander Moncrieff, R.N., the British Consul, marched from Suakin and fell prey to the Dervishes. Inexperienced and of poor calibre, the Egyptians panicked and fled leaving a third of their number dead. For Osman Digna the death of the Egyptians was unimportant. His victory lay in gaining possession of several hundred rifles and the entire extra ammunition destined for the defenders of Tokar. The effect on recruiting for Osman Digna was stupendous.

1883 had been a year of disaster for the Egyptians and 1884 would prove equally so, not only for Egypt but for Britain and imperial prestige.

The Sudan 1884

It was obvious in Cairo that this situation could not continue, for the Dervishes were now in a position to menace Khartoum itself. An army of 3,600 men with six field guns was assembled under Valentine Baker Pasha. Most of the force consisted of the 'Gendarmerie', who were described as 'a rubbishy lot of worthless ex-soldiers', by Andrew Haggard (a serving officer in the King's Own Borderers) in *Under Crescent and Star*, William Blackwood, 1896. There was a serious attempt at mass desertion when the men were ordered to

form part of the expedition. Embarking, the force sailed for Trinkitat on the Red Sea coast of the Sudan, some sixty miles south of Suakin, and well positioned for the relief of Tokar. On 4 February the army reached El-Teb only to be confronted by the Dervish forces. The ensuing action was disastrous for the Egyptians. The *Standard* war correspondent described the engagement: 'The enemy now gathered thickly and advanced towards us, and at nine o'clock showed in considerable force on some slightly rising ground, near the water springs, while on our left front I could see clumps of spears with bannerets partially concealed amidst the hillocks and bushes. Our guns again opened fire; but the shell seemed to pass over the enemy's heads ...

'Just before this, I had ridden along by the infantry column, and I saw that it was advancing in the most disorderly manner. There was no sign of discipline or steadiness; it was a mere armed mob tramping along. I was convinced they would break at the first charge. As the cavalry rode wildly in, the order was given for the infantry to form square – a manoeuvre in which they had been daily drilled for weeks. At this crisis, however, the dull, half-disciplined mass failed to accomplish it. Three sides were formed after a fashion, but on the fourth side two companies of the Alexandria Regiment, seeing the enemy coming on leaping and brandishing their spears, stood like a panic-stricken flock of sheep, and nothing could get them to move into their place. Into the gap thus left in the square the enemy poured, and at once all became panic and confusion. The troops fired indeed, but for the most part straight into the air. The miserable Egyptian soldiers refused even to defend themselves, but throwing away their rifles, flung themselves on the ground and grovelled there, screaming for mercy. No mercy was given, the Arab spearmen pouncing upon them and driving their spears through their necks or bodies. Nothing could surpass the wild confusion: camels and guns mixed up together, soldiers firing into the air, with wild Arabs, their long hair streaming behind them, darting among them, hacking and thrusting with their spears.

'While the charge had been made by the enemy on the left flank, General Baker with his Staff

Sudan 1884. The inspection of the 2nd detachment of the Guards Camel Corps at Dongola: General Salute. Note the NCOs keeping camels in line while the soldiers present arms. Sketch by Melton Prior, war artist, *Illustrated London News*. (National Army Museum)

were out with the cavalry in front. Upon riding back they found that the enemy had already got between them and the column . . . When the General finally reached the square, the enemy had already broken it up, and it was clear that all was lost.'

Any attempt on Baker's part to rally the troops was hopeless, and the army fled back to their ships leaving a trail of dead – and to the enemy, 3000 rifles, machine guns and Krupp field guns. Four days later Sinkat fell, and only six men and thirty women survived out of 400 men and numerous women and children who had attempted to leave the town.

London at last stirred itself into action and a telegram was sent to the Commander of the British Army of Occupation in Egypt, ordering him to detach a portion of his army under Sir Gerald Graham to relieve Tokar. The force, augmented by some troops on their way home from India, consisted of 2,850 infantry, 750 mounted troops, 150 Bluejackets, 100 Royal

Artillery, 80 Royal Engineers, six machine guns and eight 7-pounder guns. On their arrival at Trinkitat news was received that Tokar had fallen, but even so Graham decided to push on and engage the enemy if he could. Under a flag of truce he sent a letter to the sheiks calling on them 'to disperse your fighting men before daybreak tomorrow, or the consequences will be on your own head.' The enemy showed no signs of complying, and on 29 February the huge square formation, with transport animals in the centre, a squadron of 10th Hussars in front to scout, the rest of the cavalry in the rear and artillery and machine guns suitably positioned, moved forward over the barren sandy soil. The line of march was strewn with the remains of Baker's ill-fated expedition and swarms of carrion crows hovered over the area.

The Dervishes had entrenched themselves well and with the aid of the captured Krupp guns opened fire on the advancing square. Graham ordered his square to advance towards the enemy's left flank and by noon the formation was halted and the artillery brought into action against the Dervishes. The two field guns possessed by the enemy were silenced but small-arms fire continued to fall on the attackers. The soldiers were becoming impatient at this inactivity but Graham soon ordered the advance. 'It is not a charge,' wrote one of the war correspondents, 'but a steady solid movement in the formation which has all along been observed. It looks, however, all the more formidable, for enthusiasm and discipline are equally marked, as the whole of the troops are cheering, while the square sweeps towards the enemy.' When the square was within 200 yards of the Dervishes the enemy ceased firing and, grabbing their spears and swords, advanced with fanatical fervour. John Cameron, war correspondent of the *Standard*, described this advance. 'So hotly do the Arabs press forward that the troops pause in their steady advance. It becomes a hand-to-hand fight, the soldier meeting the Arab spear with cold steel, their favourite weapon, and beating them at it. There is not much shouting, and only a short,

sharp exclamation, a brief shout or an oath, as the soldiers engage with their foe. At this critical moment for the enemy, the Gardener guns open fire, and their leaden hail soon decides matters.'

Having won the first line, the square was halted and adjusted for the final assault against the trenches and rifle pits the enemy had dug, which with the aid of the cavalry – who had swung wide of the square and engaged the enemy – were taken after some bitter fighting.

The action had lasted three hours and had cost Graham 34 killed and 155 wounded, with Dervish casualties estimated at over 6,000. Graham's troops also recaptured the two Krupp guns, some old brass ordnance, one Gatling gun and a vast amount of rifles and ammunition. The victorious force returned to Suakin, taking with them their spoils and the surviving inhabitants of Tokar. The reinforced stronghold of Suakin was to provide an excellent base for actions against Osman

Sudan 1885. Officers and men of the Guards Camel Regiment resting. Notice the NCO serving out water from a skin and the cooks at work. (Parker Gallery)

Sudan 1885. The first view of the enemy at Abu Klea, 17 January. Note the dress of the cavalry trooper on left and the various styles worn by the officers. (National Army Museum)

Digna, and at the beginning of March 1884 a proclamation was issued calling on the rebel chiefs to surrender. This was met with a defiant reply by Osman Digna and his chiefs. On 12 March a force left Suakin and marched about eight miles to a *zareba* (a square position formed with thickets, brush and pallisades). They advanced on Tamai and bivouacked within a mile from the enemy. The troops were harassed throughout the night by a dropping fire and as dawn came up a large body of the enemy approached the square. 'This was more than British flesh and blood, however patient, could endure,' and artillery fire was ordered to disperse the Arabs.

At last the order for the advance was given and the squares, one behind the other at a distance of a thousand yards, moved over the rough ground intersected with watercourses and boulders. The enemy attacked with remarkable ferocity and soon the whole area was clouded in dense smoke from the rifle fire of both sides. 'And now, as the pressure increased, the weak points of a square formation became visible. The companies of the

Yorks and Lancaster and the Black Watch, forming the front face, swept forward against the foe; but the remaining companies of those regiments, which formed the sides of the square, and were also expecting an attack, did not keep up with the rapid movement of those in front, the consequence being that many gaps appeared in what should have been a solid wall of men.' The Dervishes took advantage of this and despite the orders of the officers and the shrill calls of the bugle which were drowned in the din, the 'Fuzzy Wuzzies' broke the square. Bennet Burleigh, the famous war correspondent of the *Daily Telegraph*, was in the square and described the scene as the Dervish swordsmen rushed into the confused mass of troops. 'The 65th (York and Lancaster) gave way, and fell back on the Marines, throwing them into disorder, though many men disdained to turn their backs, but kept their faces to the foe firing and thrusting with the bayonet. Both regiments were inextricably huddled together, and through the smoke at this dire crisis the dark demon-like figures of the foe could be seen rushing on, unchecked even for a moment by the hailstorm of bullets, and then the fight became hand-to-hand.'

Luckily, there was no panic, and the troops retired in good order, the Naval Brigade staying by their Gatlings to the last moment before locking the breech mechanism. The rear square, however, pressed forward to the right and drove everything before it, taking some of the pressure off the other troops. The 'broken' square rallied – some maintained, through the efforts of Bennet Burleigh, who above the shouting and din was heard ordering 'Men of the 65th – close up! Give it to the beggars. Let 'em have it boys! Hurrah!'

The combined fire of both squares soon decided the outcome, but it was found dangerous to move over the battlefield because of the wounded Dervishes. They accepted no quarter and demanded none. Many a British soldier to his bitter regret offered water to a wounded Arab, only to be wounded, maimed or killed by the enemy he was trying to help.

The victories at El-Teb and Tamai, however, did not prevent the Government's action in ordering a general evacuation of the Sudan.

Sudan 1885. The battle of Abu Klea, 17 January. 'Our square advancing to the attack under a very heavy fire from the enemy.' Note the zarebas made of boxes, camel saddles and thorn bush with the wounded in the centre. (National Army Museum)

Gordon, Khartoum and the Nile Expedition

While the events previously described were happening, there was another more serious occurrence which would thrust the Sudan into the limelight. The War Office and the British Government had decided to send General Charles Gordon to Khartoum. Forced to take action, Gladstone and his Government were given the choice between a costly imperial expedition to the Sudan, distasteful to the Prime Minister, or a complete evacuation of troops and civilians. To his everlasting shame, Gladstone chose the latter course of action. Although Gordon had been a popular choice at the outset, he soon grew tiresome to the Government with his sheaves of telegrams suggesting action and then countermanding the suggestion. Before embarking for the Sudan Gordon had given an interview, the first of its kind, to W. T. Stead, the controversial editor of the *Pall Mall Gazette*. Gordon's main criticism of the Government's publicised intention of evacuating the Sudan was that, 'The moment it is known that we have given up the game every man will

Sudan 1885. The battle of Abu Klea. Dervishes attacking the British square by Coulan. The Dervishes broke a British square for the second time on this occasion. (Parker Gallery)

go over to the Mahdi. All men worship the rising sun. The difficulty of evacuation will be enormously increased, if indeed, the withdrawal of our garrison is not rendered impossible.'

It was a confusing situation in the Sudan, made more so by Gordon's unpredictable nature and his reluctance to order the evacuation of civilians and troops in Khartoum. Gordon, now Governor-General of the Sudan, announced to the waiting crowds in Khartoum, who had expected more than a single man, 'I come without soldiers, but with God on my side, to redress the evils of the Sudan.' He attempted to set up an ex-slave trader named Zebehr as the ruler of the province to be evacuated, but anti-slavery feeling and the Government's natural reluctance stifled the plan. Gordon was convinced that the Mahdi could be reasoned with and that restoration of order was possible; he offered him the title of Sultan of Kordofan. After the evacuation of several hundred Egyptians the Mahdi, ignoring all Gordon's attempts at reconciliation, closed on Khartoum and on 12 March the siege commenced.

The siege, however, was far from total and Gordon's steamers continued up and down the Blue and White Niles with no interference from the Dervishes. Wounded Egyptian soldiers and women and children were transferred to Berber while Gordon set about recruiting native volunteers, reinforcing the defences and personally supervising the issue of rations. The gravity of the situation had been appreciated in Britain; from Queen Victoria, Sir Garnet Wolseley, the population and the Press, warnings, pleas, and demands were made to Gladstone and his Government to act. Sir Garnet Wolseley wrote that, 'This feeling that something should be done, like a rolling snowball, will go on increasing until the Government will be forced to adopt measures to save the Khartoum garrison . . . but if nothing is done that place will be besieged, and we shall be, in my humble opinion, faced with a war on a large scale.'

However, Gladstone did nothing but continue to assure the House and the Queen that intervention was not necessary. From Cairo the British Agent, Sir Evelyn Baring, telegraphed that 'Having sent Gordon to Khartoum, it appears to me that it is our boundless duty, both as a matter of humanity and policy, not to abandon him.' Communication with Khartoum deteriorated, and Parliament and Press were suddenly made

1 **Private, Black Watch, Egypt 1882**

2 **Private, Royal Marine Light Infantry,
   Egypt 1882**

3 **Corporal of Horse, Life Guards, Egypt 1882**

1 **Officer, Coldstream Guards, Egypt 1882**

2 **Private, Scots Guards, Egypt 1882**

3 **Private, General Post Office Rifles, Egypt 1882**

B

1 Fellah, Egyptian Army, Sudan 1883
2 Private, Camel Regiment, Sudan 1884–85
3 Rating, Naval Brigade, Sudan 1884–85

MICHAEL ROFFE

C

1 War correspondent, Sudan 1884–85

2 Fellah, 10th Sudanese Battalion, Sudan 1897

3 Private, Grenadier Guards, Sudan 1898

D

MICHAEL ROFFE

1 Trooper, 21st Lancers, Sudan 1898

2 Trooper, 21st Lancers, Sudan 1898

MICHAEL ROFFE

E

1 Sergeant-Major, 21st Lancers, Sudan 1898

2 Private, Cameron Highlanders, Sudan 1898

3 Officer, Lincolnshire Regiment, Sudan 1898

F

1 Hademdowah Warrior, 1884–98

2 Taaishi Warrior, 1884–98

MICHAEL ROFFE

G

1 and 3 Dervish Infantry, 1884–98

2 Jaidia Warrior, 1884–98

H

MICHAEL ROFFE

aware of the worsening situation, when on 1 April *The Times* published a despatch from Frank Power, their correspondent in Khartoum. Power described the 'serious reverse' suffered by a sortie from Khartoum and the continuous fighting, but above all he wrote about the hopes of those in Khartoum. 'We are daily expecting British troops. We cannot bring ourselves to believe that we are to be abandoned by the Government.' *The Times* smugly commented that the Government was largely dependent on their Khartoum correspondent for information. Wolseley continued to circulate memoranda about a relief force while other political figures tried to pressure the Government. Gladstone refused to be drawn and commented in the House that 'the debates thus constantly renewed are out of all proportion to the pressure and urgency of the question, and have the effect of offering immense obstruction to important public business.'

On 2 May 1884 Berber was captured by the Mahdi's troops, and little hope was held out for the safety of the smaller garrisons. Khartoum,

however, continued to stand, and Gordon to confuse the issue with his contradictory messages and demands for troops from Britain and Turkey. The situation was worsened by the time taken for messages to be sent from and to reach Khartoum – up to three months in some cases. The plight of the town, where Gordon had imposed rationing, printed his own paper money and awarded his own medals, was described in the last despatch from Frank Power, written on 31 July but not received in London until the end of September. 'Since March 17 no day has passed without firing, yet our losses in all at the very outside are not 700 killed. We have had a good many wounded, but as a rule the wounds are slight. Since the siege General Gordon has caused biscuit and corn to be distributed to the poor, and up to this time there has been no case of any one seriously wanting food. Everything has gone up about 3000 per cent in price, and meat is, when you can get it, 8s or 9s an ober . . . When our provisions, which we have at a stretch for two months, are eaten we must fall, nor is there any chance, with the soldiers we have,

Sudan 1885. The fierce hand-to-hand fighting involved in repulsing the enemy from the square. On-the-spot drawing by Melton Prior. (National Army Museum)

Sudan 1885. After the battle of Abu Kru or Gubat; the army preparing to start for the Nile, arranging the transport of the wounded. Note the wounded carried in litters or sitting on camels at left while war correspondent Bennet Burleigh (centre) watches the wounded being tended. General Stewart, mortally wounded during the battle, lies under a sun shade on the right. (National Army Museum)

and the great crowd of women, children, etc. of our being able to cut our way through the Arabs. We have not steamers for all, and it is only from the steamers we can meet the rebels.'

The Press continued to pressure the Government, the *Pall Mall Gazette* declaring that the Government was prepared to 'let Gordon be speared with the garrisons he was sent to save!' The dissension within the Government – the bitter controversy that raged between Gladstone and Lord Hartingdon – decided the issue. Wanting to avoid any risk of the Government falling through Hartingdon's threatened resignation, Gladstone moved a motion in the House on 5 August 1884, proposing that 'a sum not exceeding £300,000 to be granted to Her Majesty . . . to undertake operations for the relief of General Gordon should they become necessary, and to make certain preparations in respect thereof.'

Sir Garnet Wolseley arrived in Egypt on 9 September to take command of the Expeditionary Force, the same day that Colonel Stewart and Frank Power left Khartoum on a steamer with

Gordon's cypher key and detailed reports of the situation. They were murdered by Arabs some days later, after landing when the steamer grounded. Preparations were put in hand to leave Egypt, but lack of a suitable number of camels and the necessary saddles prevented any forward movement until 5 October. Even so, Wolseley's preparations were quick and thorough. The main force assembled at Wadi Halfa, just before the second cataract of the Nile, where the whale boats which Wolseley had ordered began to arrive. There was a certain amount of confusion sorting out the stores for each, and also in the camp itself. Colonel Sir William Butler later described the scene: 'Horses, camels, steam-engines, heads of departments, piles of food and forage, newspaper correspondents, sick men, Arabs and generals, seemed to be all thrown together as though the goods station of a London terminus, a couple of battalions of infantry, the War Office, and a considerable portion of Woolwich Arsenal had all been thoroughly shaken together, and then cast forth upon the desert.' Wolseley had been prompted to use whalers after his success with them during the Red River Campaign in Canada in 1870, and he even went so far as to hire Canadian boatmen. The Press was not impressed with the boats, and described them as 'that un-

Sudan 1885. After the fall of Khartoum (26 January), 'General Buller's column returning from Metemmah to Korti. Gordon Pasha's soldiers from the steamers carrying our wounded through a wadi on the road to Abu Klea.' (National Army Museum)

floatable flotilla for the Nile.' Their use also played havoc with the men's uniforms, and the rowing blistered many a hand. A correspondent wrote about the soldiers that 'Their clothes suffered very severely in places, and for want of better material, the men used to sew pieces of tin, commandeered from old commissariat boxes to their nether garments, which they very facetiously said acted as a kind of sliding board.'

The route that lay before the force was more than difficult, since much of the Nile was badly charted or completely uncharted. South of Wadi Halfa, at Dongola, General Stewart was positioned with mounted infantry and a single line battalion. Beyond Dongola lay Dervish-infested country, more cataracts, and Berber, before the force came anywhere near being within striking distance of Khartoum. Progress was slow and by December Wolseley had only reached the third cataract; time was running out for Gordon. If speed was needed in reaching Khartoum, the only way lay across the desert. At Debbah the Nile swings sharply east to Abu Hamed, then south again to Atbara before turning south-west to Khartoum. This stretch of the Nile included three cataracts and would take considerable time to cover, whereas from Korti to Khartoum over the desert was only 200 miles. By the beginning of

January 1885 the desert column had been formed, consisting of 1,800 men including the Guards Camel Regiment, artillery, cavalry, a battalion of the line and one of rifles, marines and Bluejackets. The advanced elements moved out on 30 December and were followed at the beginning of January by the main body.

On 17 January the column was attacked near the wells of Abu Klea by 10,000 Dervishes. The advancing square was experiencing difficulties in keeping its dressing, a not unusual occurrence with this type of formation but made worse on this occasion by the centre being filled with camels either carrying ammunition or allocated for use by the wounded. The camels that the soldiers had been cursing ever since their first meeting would on this occasion help to save the situation. Advancing at a sluggish pace under enemy fire, the square halted to redress the rear when '. . . the enemy suddenly started up from behind flags, advanced at a quick run, in a serrated line, headed by horsemen, and charged down with the utmost fury towards the left front corner of the square.' At a range of eighty yards the withering fire of

27

Sudan 1885. The battle of Tofrek, 22 March. A force led by the 1st Bn., the Berkshire Regiment was sent from Suakin to establish an advanced depot towards Tamai. The column was halted constructing a zareba when the Dervishes attacked. The left half of the battalion was having dinner while the right half was cutting brushwood. Because of the glare, the cavalry patrols had failed to spot the enemy until they were too close. The ensuing action, which lasted fifteen minutes, was fierce as the illustration by Charles E. Fripp shows. Eventually the enemy were beaten off leaving 2,000 dead. On 1 October 1885 the regiment was rewarded for its conduct at Tofrek by being awarded the title 'Royal'. (National Army Museum)

the Guards and mounted infantry took its deadly toll, but as the enemy halted they took ground to their right with the intention of attacking the rear left corner of the square, where cursing mounted infantrymen were trying to pull the camels inside and close the gap. The retiring skirmishers hindered any firing for the moment and once they were safely inside the square, the enemy was dangerously close. Sir Charles Bersford rushed his Gatling gun to the gaping corner, while 'Battling' Colonel Burnaby of the Royal Horse Guards, who was killed during the battle, wheeled round a company of Heavy Camel Corps and aggravated the situation. The right was now losing contact rapidly with the left, violating the most elementary rules of square formation fighting. The Dervishes poured through the gap, as Gatlings jammed,

cartridges stuck in breeches, and bayonet and sword bent.

Had the square been hollow the outcome might well have been different, but with the bunched baggage animals inside the enemy had to fight for every inch. Hand-to-hand fighting ensued as the rear ranks of the front face, endeavouring to save the situation, fired into the crush of friend, foe and animals. At last the Dervishes were driven off and as one correspondent wrote '. . . not one of the Arabs that got inside left the square alive.' Lieutenant Douglas Dawson wrote after the battle, 'I think that all present would never care to see a nearer shave . . . and it is, in my opinion, due to the fact that the two sides not immediately attacked stood their ground that the enemy retired discomfited. Had the Guards moved, none of us would have lived to tell the tale.'

On the 19th the column, now only a few miles from the Nile, was once more attacked by the Dervishes at Abu Kru, the whole affair lasting only a few minutes as the enemy was beaten off and the way was opened to the Nile; Stewart had been mortally wounded during the action, and the command now devolved on Colonel Charles Wilson, a staff officer with little experience of commanding troops in action.

On the 21st four of Gordon's steamers reached Wilson with news that the garrison, though in a sorry state, was still holding out. Wilson, however, waited for three days, as he felt his men needed rest and that defensive works should be built to protect those who were not to be embarked on the steamers for Khartoum. 'At any rate,' he later said in his defence, 'there was nothing to show . . . that a delay of a couple of days would make any difference.' Two steamers left on the 24th and arrived before Khartoum on the 28th, just those 'couple of days' too late. The Mahdi's hordes had attacked and taken the town on the 25th, after 317 days of siege, and had murdered General Charles Gordon. On 5 February a telegram was received in London stating, 'Khartoum taken by the Mahdi. General Gordon's fate uncertain.' It was received with a profound sense of shock and despair, and the Queen voiced the opinion of her subjects when she wrote in her diary, 'The Government alone is to blame.'

Sudan 1896. The Sirdar, Lord Kitchener, giving orders to his ADCs. Note the typical dress and equipment of officers of the Anglo-Egyptian army. (Wilkinson-Latham)

After Khartoum 1885-1896

While Wolseley with his two forces – the desert and river columns – continued planning to crush the Mahdi, Gladstone's Government decreed otherwise. The troops were ordered back to Egypt; and they seem to have presented a motley sight. Count Gleichen, a lieutenant in the Guards Camel Regiment, recorded that on the return march a long-promised supply of boots was encountered but they turned out to be of such small sizes that the men could not get into them. 'They were as hard as bricks, there was no grease to soften them, and the only way of using them was to slit them open at the end, and shove your toes through. As for the officers, no two had the same foot-covering: field boots, lawn-tennis shoes, garters, puttees, and boots in all stages of decay and attempted repair were worn.'

While the capture of Khartoum by the Mahdi seemed to be the end of the campaign, British forces still continued to brush with the Dervishes in the Sudan. The last battle fought by Wolseley's expeditionary force was that of Kerbekan, where the river column under General William Earle

defeated the Mahdi's troops on 10 February. Earle himself was killed during the engagement. With all British troops safely back in Egypt, the only stronghold held in the Sudan was Suakin.

Although the Mahdi died in June 1885 the fight was continued by his successor, the Khalifa. Between 1885 and 1896, when the reconquest of the Sudan was undertaken, the following actions were fought:

20 March 1885.	Dervishes defeated at Hasheen.
22 March 1885.	Dervishes defeated at Tofrek.
30 December 1885.	Dervishes defeated at Ginnis.
20 December 1888.	Dervishes defeated near Suakin.
2 July 1889.	Dervishes defeated at Arguin.
3 August 1889.	Dervishes defeated at Toski.
19 February 1891.	Dervishes defeated at Tokar.

The battle of Ginnis was significant in that it was the last engagement fought by British troops in their scarlet tunics. The campaign conducted around Suakin in 1885 was also notable for the first use of Colonial troops outside their own country. An Australian force arrived on 29 March 1885 wearing '. . . the familiar red serge coat, albeit rather strange out here, but they very soon changed into Kharkee like the rest of us.' Indian troops had been engaged during the Arabi revolt and they also took part in the 1885 Suakin campaign.

The Re-conquest of the Sudan 1896-1898

All the previous expeditions since Khartoum had achieved, according to Bennet Burleigh, was to '. . . scotch the snake whose slimy trail lies all over the Soudan, marked by the desolation of ruined villages and fields, and the whitened bones of countless human beings . . . From the beginning, Mahdism, as a social movement, has only shown potency for destruction and mischief . . . Still, there was, in the zenith of its power, always the risk that, with successful Dervish raids made north of Wady Halfa, the infection of fanaticism might spread to the confines of Lower Egypt, if not to Cairo itself.'

The situation in Egypt and the Sudan in 1896 was complicated. The Sudan, technically part of the Turkish Empire, had ceased to be ruled by Egypt in 1885 and for over ten years had been left untouched except for the affairs around Suakin. Britain could not help but see the intentions of other European powers such as France, Italy and Germany with regard to the Sudan. In 1884 she had declared a Protectorate over Somaliland and the Italians and French had swiftly followed suit. In 1896, however, the Italians, intent on further expansion, had been disastrously thrashed by the Abyssinians at Adowa, and their outpost in the Sudan, Kassala (held with British concurrence) was pressured by the Dervishes. The Italians asked the British Government to relieve this pressure by a diversionary move elsewhere, where-upon to the surprise of Egypt – who was to finance it – Britain ordered an expedition to reconquer Dongola.

Sudan 1898. Private in campaigning kit. Note the white Slade Wallace equipment and the white helmet. A khaki cover was usually worn, and this can be seen on the top step to the left of the soldier. (Wilkinson-Latham/*Navy and Army Illustrated*)

The Sirdar of the Egyptian Army was Herbert Kitchener, a Royal Engineer officer like Gordon. With little money at his disposal, and with painfully few men (only 18,000) Kitchener planned his move south. His most important ally was to be the railway system which would be built over the desert, doing away with the prime importance of following the Nile and employing boats as Wolseley had done. This masterpiece of engineering, the Sudan Military Railway, was constructed by a French-Canadian Royal Engineer officer, Eduard Girouard. It enabled troops and supplies to be moved quickly and efficiently without total reliance on camels. Railway battalions were conscripted and, considering all the natural hazards, made remarkable progress in putting down the tracks.

The invading forces moved south from Wadi Halfa and at the beginning of June headquarters was established at Akasha. The advance con-

Sudan 1898. After the battle of Atbara. Mohamed, who was taken prisoner, pictured between soldiers of one of the Sudanese battalions. Photograph autographed for Bennet Burleigh and reproduced in his account of the campaign *Sirdar and Khalifa* (1898). (Wilkinson-Latham)

tinued towards Firket, where on 7 June the Dervishes under Emir Osman Azaq were engaged and beaten. The railway sadly lagged behind and by the beginning of August had not got farther than six miles beyond Firket, when floods and cholera slowed the army. Parts of the railway were carried away, but Kitchener put the men to work day and night to re-establish his vital lifeline. By September the force had reached Kerma, pushed on to Merowe on the 26th and entered Dongola on 5 October, where after a brief encounter with the Dervishes the army found the town abandoned. The campaign had undoubtedly been a success and the combined use of the railway and suitably constructed shallow draft steamers, made in transportable sections, had proved their worth. Meanwhile in Suakin little progress had been achieved, and the reinforced garrison stagnated and suffered without contributing very much.

In 1897 the advance was pushed further up the Nile. On 7 July 1897 Major General A. Hunter and his force, having set out from Kassinger, captured Abu Hamed, and Berber was secured when the Dervishes abandoned it. At the end of the year the Italians handed Kassala back to the Egyptians (much to the chagrin of the French, whose ambitions in Africa had already caused a certain Major Marchand to commence his march from the west coast with the intention of raising the tricolour at Fashoda). Altogether 1897 had been a year of consolidation, of slow advance, of efficient preparation and of 'mopping up' the remaining Dervishes around the seaboard of the Red Sea. The most important factors of the two years of campaign had been the high state of efficiency of the Egyptian army, especially the Sudanese battalions, and the indispensable Sudanese Military Railway. Both of these were to give good accounts of themselves in the final thrust on Omdurman.

Preparations for the final smashing of the Dervishes and the re-occupation of Khartoum and the Sudan commenced in December 1897. The chief events leading up to the battle of Atbara and finally Omdurman were:

Sirdar asks for reinforcements of British troops . . .	31 Dec. 1897
British brigade starts for front from Abu Dis	26 Feb. 1898
British brigade reaches Dibeika, beyond Berber	3 March 1898
Sirdar leaves Berber	15 March 1898
Concentration at Kenur	16 March 1898
Army moves up the Atbara	20 March 1898
First contact with Dervish cavalry	21 March 1898
Shendi raided and destroyed	27 March 1898
General Hunter reconnoitres Mahmud's zariba	30 March 1898
Second reconnaissance: cavalry action before Mahmud's zariba	4 April 1898

On 8 April 1898 the Anglo-Egyptian army was face to face with a large Dervish force under Mahmud, one of the Khalifa's most able generals. Kitchener had been undecided about whether to assault Mahmud's zariba but at last ordered a night march and dawn attack. The squares moved forward, stopped to rest and readjust and then moved on again, until 4 a.m. when a halt

Sudan 1898. Dervishes in full panoply mounted on camels and armed with their long-bladed double edged swords and broad-headed spears. Note the various patches sewn on the 'jibbah'. (Wilkinson-Latham/*Navy and Army Illustrated*)

Sudan 1898. A scouting patrol of the Egyptian Camel Corps. Note the khaki uniforms, blue puttees, brown leather bandoliers and Martini-Henry rifles with long bladed socket bayonet. (Wilkinson-Latham/*Navy and Army Illustrated*)

was called and the soldiers lay down and slept if they could. The enemy's position was a circular zariba behind which was the dry bed of the river Atbara. Inside the zariba facing the Anglo-Egyptian force was a stockade and trenches, with the centre composed of rifle pits and more trenches. It was this centre which proved the most difficult to clear and capture.

As the dawn rose, the men rose too and the four brigades advanced towards the enemy zariba. As G. W. Steevens recorded in his book *With Kitchener to Khartoum*: 'The word came, and the men sprang up. The squares shifted into the fighting formations: at one impulse, in one superb sweep, near 12,000 men moved forward towards the enemy. All England and all Egypt, and the flower of the black lands beyond, Birmingham and the West Highlands, the half-regenerated children of the earth's earliest civilisation, and grinning savages from the uttermost swamps of Equatoria, muscle and machinery, lord and larrikin, Balliol and the Board school, the Sirdar's brain and the camel's back – all welded into one, the awful war machine went forward into action . . . The line went on over the crunching

gravel in awful silence, or speaking briefly in half-voices – went on till it was not half a mile from the flags. Then it halted. Thud! went the first gun, and phutt! came faintly back, as its shell burst on the zariba into a wreathed round cloud of just the zariba's smoky grey.'

After an hour and twenty minutes of intensive bombardment the guns fell silent, much to the relief of the soldiers' ears, and the word to advance was given. The Cameron Highlanders were first to the zariba and, tearing aside the low loose hedge of camel thorn, were over the stockade and into the trenches. The interior of the zariba, filled with mimosa thorn and grass, was honeycombed with rifle pits and holes. The troops surged forward killing every Dervish in their path. G. W. Steevens witnessed the scene: 'Bullet and bayonet and butt, the whirlwind of Highlanders swept over. And by this time the Lincolns were in on the right, and the Maxims, galloping right up to the stockade, had withered the left, and the Warwicks, the enemy's cavalry definitely gone, were volleying off the blacks as your beard comes off under a keen razor. Farther and farther they cleared the ground – cleared it of everything like a living man, for it was left carpeted thick enough with dead. Here was a trench; bayonet that man. Here a little straw tukl; warily round to the door, and then a volley. Now in column through this opening in the bushes; then into line, and drop those few desperately firing shadows among the dry stems beyond. For the running blacks – poor heroes – still fired, though every second they fired less and ran more. And on, on the British stumbled and slew, till suddenly there was unbroken blue overhead, and a clear drop underfoot. The river! And across the trickle of water the quarter-mile of dry sand-bed was a fly-paper with scrambling spots of black. The pursuers thronged the bank in double line, and in two minutes the paper was still black-spotted, only the spots scrambled no more. "Now that," panted the most pessimistic senior captain in the brigade – "now I call that a very good fight."' The entire action after the bombardment had lasted forty minutes and had cost the Anglo-Egyptian force 81 killed and 493 wounded.

Sudan 1898. The first sight of the dervish horde at Omdurman. British and (extreme right) Egyptian army officers watching the enemy movements. (National Army Museum)

The calendar of events after the battle was as follows:

Sirdar's triumphal entry into Berber	11 April 1898
Railhead reaches Abeidieh: construction of new gunboats begun	18 April 1898
Railhead reaches Fort Atbara	(middle) June 1898
Lewis's Brigade leaves Atbara for south	(early) July 1898
Second British brigade arrives at Atbara	3–17 August 1898
Sirdar leaves Atbara for front	13 August 1898
Last troops leave Atbara	18 August 1898
Final concentration at Gebel Royan	28 August 1898
March from Gebel Royan to Wady Abid (eight miles)	29 August 1898
March from Wady Abid to Sayal (ten miles)	30 August 1898
March from Sayal to Wady Suetne (eight miles)	31 August 1898
Kerreri reconnoitred and shelled	31 August 1898
March from Wady Suetne to Agaiga (six miles)	
Omdurman reconnoitred and forts silenced	1 September 1898

As dawn rose on 2 September, the Sirdar and his Anglo-Egyptian force of about 25,000 men were assembled in a horseshoe formation, with each flank touching the Nile where suitable pro-

No............. To ....Col Martin.....
At ....
From ....
At ....
Hour... 8. 30..... Day... 2.. .9...18 98

Annoy them as far as possible on their flank, head them off if possible from Omdurman

Signature ....Sirdar

Sudan 1898. The order that resulted in the famous charge of the 21st Lancers at Omdurman. Ranking in spectacle with the charge of the Light Brigade in the Crimea, this charge was considered by many as 'indisputable folly'. (Wallis and Wallis/R. Butler)

Sudan 1898. The charge of the 21st Lancers at Omdurman. A colourful impression of the action by the military artist R. Caton Woodville. (Parker Gallery)

tection was given by shallow draft gun boats. The position was only seven miles from Omdurman. In front of the army lay a dry open plain dotted with grass, devoid of cover except for several dried-up water courses and a few folds in the terrain. At dawn the cavalry, both British and Egyptian, had gone out but at about 6.30 a.m. they came in. 'The noise of something began to creep in upon us' wrote Steevens, 'it cleared and divided into the tap of drums and the far-away surf of raucous war cries . . . They were coming on.' On the right was seen the black banner of the Khalifa's brother and on the left the blue and white banners of his son. 'They came very fast, and they came very straight' continued Steevens, 'and then presently they came no farther. With a crash the bullets leaped out of the British rifles . . . section volleys at 2,000 yards. . . . The British stood up in a double rank behind their zariba; the blacks lay down in their shelter-trench; both poured out death as fast as they could load and press trigger.' The Dervishes did not stand a chance but evoked the admiration of those who witnessed the fight. 'And the enemy? No white troops would have faced that torrent of death for five minutes, but the

Baggara and the blacks came on. The torrent swept into them and hurled them down in whole companies. You saw a rigid line gather itself up and rush on evenly; then before a shrapnel shell or a Maxim the line suddenly quivered and stopped. The line was yet unbroken, but it was quite still. But other lines gathered up again, again, and yet again; they went down, and yet others rushed on. Sometimes they came near to see single figures quite plainly . . . It was the last day of Mahdism, and the greatest. They could never get near, and they refused to hold back. By now the ground before us was all white with dead men's drapery. Rifles grew red-hot; the soldiers seized them by the slings and dragged them back to the reserve to change for cool ones. It was not a battle, but an execution.'

Kitchener, seeing the effect of this murderous fire which, as Bennet Burleigh wrote, 'was reaping a gigantic harvest', knew that if the Dervishes were able to get amongst his lines his weapon superiority would be useless. He quickly and surprisingly ordered a counter-attack which, largely due to the Sudanese troops, sealed the Khalifa's fate. The brigades advanced in perfect order and

34

the troops were able to see '. . . what awful slaughter we had done. The bodies were not in heaps – bodies hardly ever are; but they were spread evenly over acres and acres.' The enemy were not all dead, some feigning death or slightly wounded. Steevens recorded that, 'Some lay very composed with their slippers placed under their heads for a last pillow; some knelt, cut short in the middle of a last prayer. Others were torn to pieces, vermilion blood already drying on brown skin, killed instantly beyond doubt. Others again, seemingly as dead as these, sprang up as we approached and rushed savagely, hurling spears at the nearest enemy. They were bayoneted or shot.'

At 8.30 a.m. the Sirdar had instructed Colonel Martin of the 21st Lancers (a regiment given the unofficial motto 'Thou shalt not kill' by the rest of the army, because they possessed no battle honours) to annoy the Dervishes on their flank and to head them off from Omdurman. The regiment moved off towards Omdurman and shortly afterwards came across some 300 Dervishes. To cut them off, it was thought better to go a little west, wheel and gallop down on them. 'The trumpets sang out the order' wrote Steevens, 'the troops glided into squadrons, and, four squadrons in line, the 21st Lancers swung into their first charge. Knee to knee they swept on till they were but 200 yards from the enemy. Then suddenly – then in a flash – they saw the trap. Between them and the 300 there yawned suddenly a deep ravine; out of the ravine there sprang instantly a cloud of dark heads and a brandished lightning of swords, and a thunder of savage voices . . .'

'Three thousand, if there was one, to a short four hundred; but it was too late to check now. Must go through with it now! The blunders of British cavalry are the fertile seed of British glory: knee to knee the Lancers whirled on. One hundred yards – fifty – knee to knee – Slap! "It was just like that," said a captain, bringing his fish hard into his open palm. Through the swordsmen they shore without checking – and then came the khor. The colonel at their head, riding straight through everything without sword or revolver drawn, found his horse on its head, and the swords swooping about his own. He got the charger up again, and rode on straight, unarmed, through everything. The squadrons followed him down

the fall. Horses plunged, blundered, recovered, fell; Dervishes lay on the ground for the hamstringing cut; officers pistolled them in passing over, as one drops a stone into a bucket; troopers thrust till lances broke, then cut; everybody went on straight, through everything.'

Having passed through the enemy, the lancers dismounted and fired at the Dervishes with their carbines, driving them back towards the artillery. 'The shrapnel flew shrieking over them,' wrote Steevens, 'the 3000 fell all ways and died.' A first-hand account of the charge was graphically written for the readers of the *Morning Post* by their correspondent, Winston S. Churchill, 4th Hussars, attached to the 21st Lancers.

Although the charge was costly, and almost – as the headlines stated – 'A Second Balaclava charge', the Dervishes were cut off from Omdurman and at 11.30 a.m. the battle was virtually over. By midday victory was complete, and at the beginning of the afternoon Omdurman was in the hands of the Sirdar's troops. The final Dervish resistance was crushed and the field occupied as the advancing battalions cleared the city and the area around it. 'The last Dervish stood up and filled his chest; he shouted the name of his God and hurled his spear. Then he stood quite still, waiting. It took him full; he quivered, gave at the knees, and toppled with his head on his arms and his face towards the legions of his conquerors.'

'Sir Herbert Kitchener's Brilliant Victory' screamed the newspaper headlines, but while the Dervishes had been 'killed out as hardly an army had been killed out in the history of war,' many thought it not all due to the Sirdar's brilliance. The battle was, as Steevens wrote, '. . . a miracle of success. For that thanks are due to the Khalifa, whose generalship throughout was a masterpiece of imbecility . . . the Sirdar would have won in any case; that he won so crushingly and so cheaply was the gift of luck and the Khalifa.' The charge of the Lancers, the heroic episode in which three Victoria Crosses were won, also came in for its share of rebuke; 'For cavalry to charge unbroken infantry, of unknown strength, over unknown ground . . . was as grave a tactical crime as cavalry could possibly commit.' The regiment had, however, earned its first battle honour, 'Khartoum'.

Sudan 1898. Private T. Byrne, 21st Lancers. One of the three members of the regiment to gain the Victoria Cross for his heroic action during the charge. Pictured in Cairo after the campaign, he wears the uniform worn on that day. Note the wide sun shade to the helmet, furled lance pennon, picket and spare boots on the holsters. (Wilkinson-Latham/*Navy and Army Illustrated*)

The fight in the Sudan was almost over, but not quite. Various Dervishes still had to be subdued and this was achieved on 22 September at Gedaref in the Eastern Sudan, against Ahmed Fedil. The following year Ahmed Fedil, having escaped and rejoined the Khalifa, who was still at large, was killed together with his master on 22 November 1899. Osman Digna was captured in January 1900 and lived until 1926. The reconquest of the Sudan was complete and Gordon revenged.

Between 1900 and 1908 the Sudan still continued to occupy the Egyptian army and various small insurrections had to be dealt with, either by police action or punitive expeditions.

'The poor Sudan! The wretched dry Sudan!' wrote Steevens at the end of the 1898 campaign, 'Count up all the gains you will, yet what a hideous irony it remains, this fight of half a generation for such an emptiness.'

In 1956 Britain finally withdrew from Egypt and the 'emptiness' that had been named the Anglo-Egyptian Sudan.

The casualties of the enemy were enormous: 9,700 dead and an estimated 10,000 wounded, and 4000 taken prisoner. The first estimation of the Anglo-Egyptian casualties, not including those who died later of wounds, was 131 British and 256 native killed and wounded.

Before the battle Kitchener had received sealed secret orders from London not to be opened until Khartoum was captured. These concerned the French Major Marchand who was now installed under the tricolour at Fashoda. Kitchener, acting on his orders to proceed upstream and dislodge any French force he found, started out on 9 September after having conducted a funeral service for Gordon and ordered the Mahdi's tomb to be destroyed. On 24 September, having confronted the French, who stood down and retired from the Sudan, Kitchener made a triumphal return to Khartoum.

The Plates

A1 Private, Black Watch, Egypt 1882
There was very little difference between the dress of the British soldier in England or in Egypt at this period except for the exchange of full dress head-dress for a white sun helmet with brass chin chain. In the Black Watch, the treasured red hackle was worn in the puggaree on the left side. The 'frock' of scarlet cloth had a blue collar with regimental badge and the letters 'R.H.' (Royal Highlanders) on the shoulder straps; it was rounded at the front to clear the sporran, with two lower pockets with three buttons and white worsted loops. The equipment was the 'Valise' pattern and worn with haversack and waterbottle on campaign.

A2 Private, Royal Marine Light Infantry, 1882
The Royal Marine Light Infantry uniform worn in Egypt in 1882 consisted of the customary white

helmet and a blue 'frock' with brass buttons. The blue collar was decorated with a bugle-horn badge in scarlet worsted and the shoulders were decorated with scarlet worsted cords. The trousers were blue with a scarlet welt on the outside of the leg and worn tucked into black leather gaiters. The equipment was the 'Valise' pattern worn with black leather pouches, which had been replaced by white in most regiments by this date. At Tel-el-Kebir *The Times* war correspondent remembered seeing the '. . . endless line of the dreaded redcoats, broken by the even *more fearful blue* of the Marines.'

A3 Corporal of Horse, Life Guards, Egypt 1882

The white helmet was worn with an unlined scarlet serge 'frock' with five brass buttons. The Corporal of Horse depicted wears a white waistbelt with sword slings to suspend the special Household Cavalry pattern sword on the left side. The waistbelt also supported a brown leather holster on the left hip with a pouch on the right side. Troopers wore the same dress but without the revolver, the pouch carrying carbine ammunition. Breeches were in blue and had a double

Sudan 1898. Captain P. A. Kenna, 21st Lancers. One of three members of the regiment to gain the Victoria Cross for his part in the charge at Omdurman. Note the officer carrying a lance and wearing a different pattern of helmet from that of the other ranks. The personal and horse equipment can clearly be seen. (Wilkinson-Latham/*Navy and Army Illustrated*)

scarlet stripe with piping between on the outer seam of each leg. Haversack and water bottle were carried on straps crossing the chest.

B1 Officer, Coldstream Guards, Egypt 1882

The officer of the Coldstream Guards depicted is wearing the dress as worn on arrival in Egypt. On the way out helmets had been stained brown, but the regimental badge was still fitted to the puggaree at the front. The Scots Guards also wore their badge, but the Grenadiers dispensed with theirs. The scarlet 'frock' had the buttons arranged in pairs, with two pockets below the waist and two patch pockets on the breast. Ranking was shown on the scarlet shoulder straps and the regiment was distinguished by the badge on the blue collar. Officers wore drab breeches and brown riding boots, or blue trousers tucked into gaiters like those of the men. The equipment was the 'Sam Browne' belt with revolver holster, pouch, sword frog and braces.

B2 Private, Scots Guards, Egypt 1882

The private shown wears essentially the same uniform as used by the line infantry involved in the campaign. This consisted of the white helmet, sometimes dyed brown as in this case (Scots Guards also wore their badge); scarlet 'frock' with blue collar (Royal regiments had blue collars, English regiments white, Scottish regiments yellow, and Irish regiments green) and brass buttons. The trousers were dark blue with scarlet welt, tucked into black gaiters. The equipment consisting of waist belt, braces, pouches, and valise, was of the 'Valise' pattern. An off-white haversack and a waterbottle were also carried.

B3 Private, General Post Office Rifles, Egypt 1882

The 24th Middlesex Rifle Volunteers, composed mainly of workers from the Post Office, furnished Sir Garnet Wolseley's Egyptian expedition with a Telegraphic Detachment. They wore their usual tunic, blue with green collar and cuffs, and dark trousers tucked into black gaiters. The 'Valise' equipment was in black leather as was usual with Rifles both regular and volunteer, and the Home pattern helmet was replaced by a white one. The campaign won for the regiment the first Volunteer overseas battle honour, 'Egypt 1881'.

Sudan 1898. The end of Mahdism: the dead Yakub and his followers beside the Khalifa's black flag. Drawing by war artist H. C. Seppings Wright. (National Army Museum/ *Illustrated London News*)

C1 Fellah, Egyptian Army, Sudan 1883

The Egyptian Army formed after the smashing of the Arabi revolt were dressed in the same fashion as the old army. Originally the new army was formed of 6,000 men with twenty-five British officers. The uniform consisted of the tarbush, a white tunic and loose trousers tucked into white gaiters. The equipment, comprising waistbelt, ammunition pouch and bayonet frog, was in black leather. A large pack was worn on the back with blankets and rolled greatcoat. The rifle was at first the American Remington rolling-block pattern with brass-hilted sword bayonet.

C2 Private, Camel Regiment, Sudan 1884–85

The Camel Regiment in the relief force sent to rescue General Gordon from Khartoum consisted of detachments from the Guards, Household Cavalry, regular cavalry and Royal Marines, their task being to act as mounted infantry. The uniform was the white helmet with brass chin chain, blue goggles against glare from the sun and a green net against flies. The 'frock' was grey with two patch pockets and five brass buttons down the front. Drab Bedford Cord breeches were worn with dark blue puttees and brown boots. The equipment consisted of the waistbelt and pouch of the 'Valise' equipment, a brown leather bandolier, haversack and 'Oliver' pattern waterbottle. The Martini-Henry rifle was carried with the long 'yatagan'-bladed sword bayonet.

C3 Rating, Naval Brigade, Sudan 1884–85

There were a number of 'Bluejackets' landed from ships with the Khartoum relief force, mainly employed in manning the Gatling guns. The uniform was the standard blue blouse and bell-bottom trousers tucked into black gaiters, and the head dress was the sennet hat in straw with the name of the ship on a blue 'tally' band. On occasion the small round hat with white cover was also worn. Equipment consisted of brown leather belt, pouches, and braces, haversack and waterbottle. A grey blanket was carried, rolled and worn over the left shoulder, with the ends tied together on the right hip. Ratings carried the cutlass bayonet for the Martini-Henry rifle.

D1 War Correspondent, Sudan 1884–85

The figure represents the typical dress worn by the war correspondents, although individual tastes were often apparent. St. Leger Herbert of the *Morning Post*, killed at Abu Kru in 1885, wore a red tunic; others wore grey or drab. The figure (representing Bennet Burleigh of the *Daily Telegraph*) wears a blue serge jacket, dyed helmet, brown leather waistbelt with revolver holster and pouch, a binocular case slung over the shoulder and usually a brown leather note case with pad and pencils. Breeches were drab and worn with either boots or puttees.

D2 Fellah, 10th Sudanese Battalion, Sudan 1897

The standard uniform of the Egyptian army had altered from the previous pattern shown and was now the tarbush, a brown jersey, sand coloured trousers and dark blue puttees. The equipment was in brown leather and consisted of belt, pouches and braces and an additional bandolier for ammunition. Haversack and waterbottle were carried, and a blanket when needed. Egyptian troops wore a neckflap to their khaki tarbush cover, but the Sudanese battalions dispensed with this. Each battalion bore its number on a coloured cloth patch on the right side.

D3 Private, Grenadier Guards, Sudan 1898

The infantry of the line and the Guards were barely distinguishable in the khaki drill foreign service dress they wore in the Sudan, except for the flash on the helmet and the shoulder strap badges. The helmet had a khaki cover and neck flap and a brown leather chin strap. In the case of the Grenadier Guards, both officers and men wore a grenade-type design in black cloth. Line regiments identified their helmets with pieces of the shoulder straps of the scarlet 'frock', which bore the regimental title in white worsted embroidery; Scottish regiments, however, tended to use a patch of their kilt tartan. Tunic and trousers were in khaki drill, the former with two patch pockets and the latter tucked into khaki puttees. The equipment was the 1888 pattern Slade-Wallace in white buff leather, which was usually left unclean on campaign.

E1 Trooper, 21st Lancers, Sudan 1898

This and the next trooper wear the same uniform, this rear view showing the chain shoulder straps, the spine pad and the method of hooking up the 1890 pattern sword with its steel scabbard covered in khaki cloth. The quilted sun shade to the helmet is also shown, in this instance tucked under the puggaree. There were several methods of fixing the sun shade, contemporary photographs as well as the *Illustrated London News* and *Graphic* showing various combinations.

E2 Trooper, 21st Lancers, Sudan 1898

The figure represents Trooper Byrne V.C., a participant in the famous charge at Omdurman. The cavalry wore the same pattern tunic as the infantry but with breeches and puttees. There was no waist belt over the tunic, the sword belt being worn under. When mounted the sword was carried on the saddle in the 'shoe case'. A bandolier in brown leather was worn over the left shoulder with ammunition for the carbine, which was carried on the right rear of the saddle in a brown leather holster. Various other items such as

Sudan 1898. Looters scouring the field after the battle in search of weapons and valuables. This practice on the part of Egyptian troops was heavily criticised in the press, as was the wholesale killing of wounded Dervishes on the field. (National Army Museum)

pickets and forage were carried by cavalry on campaign, and being lancers the rank and file of the regiment were armed with a lance with steel head and butt, decorated with a red-over-white pennon. This was furled, as shown, on active service.

F1 Sergeant-Major, 21st Lancers, Sudan 1898

The Sergeant-Major had a uniform akin to that of the officers. He wore the standard khaki helmet and neck flap but with a tunic like the officers', single-breasted with four patch pockets and two buttons on the sleeve. The breeches were the same as the rank and file, drab Bedford cord with leather reinforce tucked into brown leather gaiters and brown boots. Equipment was the 'Sam Browne' belt in brown leather with braces, pouch, sword frog and holster for the Webley Revolver. The badge of rank, a crown on the lower sleeve, was in gold embroidery.

F2 Private, Cameron Highlanders, Sudan 1898

The figure is taken from the well-known sketch by W. T. Maude, war artist during the Omdurman campaign for the *Graphic* newspaper. It portrays a private of the regiment running back from the firing line with overheated rifles, to exchange them with those of the reserve line. The uniform

Sudan 1898. Seaforth and Cameron Highlanders burying the dead after the battle of Omdurman. Casualties in the British and Egyptian armies were relatively light compared with the slaughter of the Dervishes, some 10,000 of whom lay dead on the field. (National Army Museum)

was typical of that worn by Highland regiments during this campaign. The khaki helmet with tartan flash had a brown leather chin strap, and the tunic had the front rounded to clear the sporran. On occasion the standard infantry pattern tunic was issued to Highland regiments in lieu of the pattern shown. The equipment was the Slade Wallace pattern, and the rifle the Lee Enfield with the bayonet 1888 Mk. II.

F3 Officer, Lincolnshire Regiment, Sudan 1898

The officer shown is dressed in the typical uniform of the infantry officer during the campaign. The khaki helmet was a different shape from that of the rank and file, and was later known as the Wolseley pattern. The tightly wound puggaree on the helmet had the regimental identification on the right side. Although orders stated that officers should wear the badge as used on the field service cap, they tended to utilise the cut-down 'frock' shoulder straps of the rank and file, Scottish regiments excepted. The khaki tunic had four patch pockets and was worn with breeches and brown leather gaiters, although some officers preferred puttees or riding boots. The equipment was the usual 'Sam Browne' belt, with braces, pouch, revolver holster and sword frog. The 1895 or 1897 pattern infantry sword was carried in a brown leather scabbard with steel chape. Scottish regiments carried the 'claymore', Guards and Rifles swords of their own patterns.

G and H Dervishes, Sudan 1884–98

The dress of the Dervishes varied according to the tribe and the area from which they came, but all wore the basic *jibbah*, a loose short blouse in white cotton (some with blue stripes) patched with squares of black, red, blue and yellow cloth in imitation of the Mahdi, who wore this garb to emphasise poverty. The word 'Dervish', from the Persian 'darvesh', was an all-embracing word meaning poor. The original *jibbahs* were white, with the coloured cloth patches applied to mend wear and tear, but later these patches were used as symbolic badges of the *Ansars* (helpers) of the Mahdi. The figures show the various types of troops in the Mahdi's and later the Khalifa's army. *G1* shows a Taaishi warrior of the Baggara tribe mounted on a camel. He wears the *jibbah* with patches and carries a sharp double-edged sword in a leaf-shaped scabbard, also spears. E. N. Bennet, the war correspondent of the *Westminster Gazette*, noted that 'The cross handled Dervish sword is terribly heavy and the long straight blades . . . freshly ground . . . The large Dervish spear, too, when properly handled, is a most formidable weapon, and if a thrust is driven well home into the body, the wound from the broad iron head is so wide and deep that a man has little chance of recovery.' *G2* shows a Hadendowah, one of the fiercest tribes in Africa according to some, whose hair spawned the name of 'fuzzy wuzzy'. They wore cotton trousers patched like the *jibbah* and carried the sharp double-bladed sword and hide shield. One of their habits was to attack the enemy's hands first, hacking them off. In battle they would also lie still, feigning dead, and then hamstring horses or hack at the hands of British or Egyptian troops. The other figures show Dervishes in various dress: *H3* shows a black Jiadia rifleman armed with a Remington captured from the massacred Hicks column. The Dervishes understood little about rifles, and according to E. N. Bennet knocked the sights off and cut down the barrels to suit themselves, with dire consequences to accuracy. The opening volley at Omdurman was recorded by many war correspondents as being a simple waste of ammunition on the part of the enemy.